How to Grow Marijuana

~

A Beginners Guide to Indoors/Outdoors Growing for Medicinal use + Recipes

By

Josh Newman

Table of Contents

ISBN-13: 978-1547145201
ISBN-10: 154714520X

Introduction

Medicinal cannabis is used to treat diseases and relieve symptoms covering a whole gambit of afflictions. It can be smoked, ingested in food, and taken as a pill or as an oil.

The story of medical cannabis use in the United States is fraught, complicated, and ongoing but it does serve as a useful tool in managing many people's health needs. Luckily, laws in the United States are gradually acknowledging this.

Today, we'll start off by discussing some of the history of medical cannabis and the laws regulating it state by state.

We'll also give a brief overview of the basics of the plant's anatomy and then move into the supplies and best practices that will help you grow successfully.

By the time you finish reading, you'll have a solid understanding to begin the rewarding process of growing medicinal cannabis for yourself.

Let's get started!

Medical Uses of Cannabis

Cannabis has been used to provide sufferers with relief from a wide variety of symptoms and illnesses. Here's a brief list of conditions that have benefitted from the medical use of cannabis:

- Chronic pain
- Nausea and vomiting especially due to cancer treatments
- Wasting syndrome
- Crohn's disease
- Epilepsy
- Amyotrophic lateral sclerosis (ALS)
- Anorexia due to HIV/AIDS
- Multiple sclerosis related muscle stiffness or severe muscle spasms
- Tourette syndrome
- Alzheimer's disease
- Terminal illness

Side Effects and Effectiveness
Cannabinoids feature many different chemicals but the two linked to positive side effects for medical treatment are tetrahydrocannabinol (THC) and cannabidiol (CBD). Currently, the FDA has approved two drugs- Dronabinol and Nabilone- that include ingredients found in cannabis.

Like with most medications, there is some variability in terms of side effects and level of relief from symptoms. Some patients may experience cannabis taking effect in as little as 30 minutes or as long as a few hours. The timing of its effects are thought to be harder to control when taking cannabis in pill form rather than through smoking. It is also thought that after ingesting cannabis, it typically takes an hour for the effects to manifest whereas with smoking the effects can be felt almost immediately.

There are some forms of medicinal cannabis designed to provide relief without the intoxicating or mood-altering effects often associated with recreational use. This is especially useful when treating children. For more on this, it is best to consult your own physician.

ANNUAL DEATHS

Tobacco	435,000
Poor Diet/Exercise	365,000
Alcohol	85,000
Prescription Drugs	32,000
Motor Vehicle Crashes	26,347
Homicide	20,308
Aspirin	7,600
Peanuts	100
Marijuana	0

MARIJUANA: Safer Than Peanuts!

History and Laws in the US

First off, it's important to acknowledge that marijuana, pot, weed, and ganja are all slang terms for the cannabis plant. The use of the word "marijuana" was promoted by prohibition supporters in the 1920s and 1930s along with the nickname "devil's weed". Using a Spanish word helped to create a connection between the anti-immigration movement and the anti-drug movement as many thought that Mexican immigrants brought cannabis into American culture. Cannabis is the correct term for describing this multi-use plant and its byproducts and it is the term that will be used throughout this book.

Secondly, cannabis use has been legalized in 29 states but each state has their own conditions of use that you need to be aware of if you have been prescribed cannabis for medicinal use. Generally, a few aspects of the process are true almost universally in state with some sort of medicinal cannabis program and legislature. For example, you need to be an adult which means age 21 or older, have a prescription from a physician, and hold a government issued form of ID.

Oftentimes, you need to be registered to your state's Medical Marijuana Registry and hold a card stating you are on that list as well. Within most state systems, a patient can also designate a caregiver, someone who can cultivate and transport cannabis for the patient if the patient is a child or the patient cannot procure the cannabis for themselves due to medical or travel restrictions. The caregiver in most cases will also get their own card or number stating that they are registered to be a caregiver.

The details below were compiled on May 9th 2017 and break down each state's regulations for possessing and growing as of that date.

Alaska

Possession & Use	Grow Limits
Adults (age 21+) may display, purchase, transport, possess, and use at the most 1oz of cannabis.	Adults (age 21+) may grow, process, possess and transport a maximum of 6 plants total, with at most 3 plants that are mature.

Arizona

Possession & Use	Grow Limits
A patient or registered caregiver is allowed to obtain and possess at most 2.5oz of cannabis in a period of 14 days from a registered nonprofit medicinal cannabis dispensary.	A patient or caregiver may grow up to 12 plants if they live more than 25 miles from the closest dispensary.

California

Possession & Use	Grow Limits
Adults (age 21+) may possess any combination of: • 28.5oz usable cannabis. • 8g cannabis concentrate. Consumers may only purchase 8oz per day.	Adults (age 21+) may grow at most 6 plants per residence. Patients or their caregivers may grow at most 6 mature cannabis plants or at most 12 immature cannabis plants. If recommended by a physician, patients may grow more plants to fit the patient's needs.

Connecticut

Possession & Use	Grow Limits
A patient and caregiver's combined amount of useable cannabis cannot exceed an amount reasonably necessary to ensure an uninterrupted supply for 1 month.	Home growing is not allowed.

Colorado

Possession & Use	Grow Limits
Adult (age 21+) residents of Colorado may purchase and possess at most 1oz of cannabis at any time. Adult non-Colorado residents may only purchase or possess 1/4 oz. Medical use of cannabis is legal as long as: • A patient possesses no more than 2oz of a useable form of cannabis. • Possesses at the most six plants with 3 or less flowering and producing a useable form of cannabis at one time.	Adult (age 21+) residents of Colorado may grow 6 plants per person, but at most 3 plants can be in the flowering stage at any time. Non-Colorado residents are not permitted to grow cannabis. There are at most 12 total plants allowed per residence. The number of adults living there is inconsequential. Plants must be kept out of view and in an enclosed, locked area. Cannabis that is home grown may not be sold.

Delaware

Possession & Use	Grow Limits
6oz is deemed a useable amount of cannabis for medical use.	Home growing is not allowed.

Washington D.C.

Possession & Use	Grow Limits
Adults (age 21+) may possess at the most 2oz of cannabis. It is permitted to transfer 1oz or less of cannabis to another person who is at least 21 years old as long as there is no payment or a exchange of goods or services. Adults may also possess cannabis related drug paraphernalia that accommodates 1oz of cannabis or less. Cannabis may be used on private property. The maximum amount any patient or caregiver may possess at any time is 2oz or the equivalent of 2oz of dried medical cannabis when sold in another form.	An adult (age 21+) may grow up to 6 cannabis plants within their residence as long as no more than 3 are mature.

Florida

Possession & Use	Grow Limits
Specific regulations have yet to be determined.	Home growing is not allowed.

Hawaii

Possession & Use	Grow Limits
The amount of cannabis possessed by a patient cannot exceed an adequate supply which is deemed by a physician.	A medicinal cannabis patient may grow at most 7 plants total and there can be no more than 4oz of useable cannabis between a patient and a caregiver. A designated caregiver may grow no more than 7 plants total and may only grow for one patient at a time. Each plant should be tagged at the base with their registration card number and expiration date. After 12/31/18 a caregiver may not grow cannabis on behalf of a patient unless the patient is a minor, an adult lacking in legal capacity, or lives on an island without a dispensary.

Illinois

Possession & Use	Grow Limits
An adequate supply is legally considered to be 2.5oz of usable cannabis during a period of 14 days. It must be derived from a state source	Home growing is not allowed.

Maine

Possession & Use	Grow Limits
An patient or adult (age 21+) may possess at most 2.5oz of prepared cannabis, 2.5oz of cannabis concentrate, or 2.5oz of cannabis infused edible products.	A Maine resident may have at most 6 plants for personal use. The plants must be kept in an enclosed, locked, area unless they are being transported. Minors, homeless patients, patients in hospice or nursing facilities, and incapacitated adults may not cultivate their own cannabis. Only primary caregivers or dispensaries may cultivate on behalf of a patient in these specific cases.

Massachusetts

Possession & Use	Grow Limits
Adults (age 21+) may possess in any combination: • A maximum of 1oz of usable cannabis in a public setting. • A maximum of 10oz of usable cannabis in a private setting. • A maximum of 5g of cannabis concentrate.	Adults (age 21+) may grow up to 6 mature plants per person. The plants must not be visible to the public. Patients with hardships pertaining to cultivation registration may grow enough plants to maintain a 60-day supply for that patient's use. Patients may apply for a hardship cultivation registration if they can demonstrate that their access to a dispensary is limited by: • Financial hardships. • Physical incapacity barring access transportation. • Lack of a dispensary within a reasonable distance of the patient's residence and lack of a dispensary that will deliver cannabis to the patient's address.

Michigan

Possession & Use	Grow Limits
A patient who has been issued and possesses a registry identification card as well as a state identification card will not be subject to arrest or penalty in any manner provided that the patient possesses an amount of useable cannabis that does not exceed 2.5oz.	A patient or primary caregiver cannot have more than 12 plants and they must be housed in an enclosed and locked facility.

Montana

Possession & Use	Grow Limits
A patient may possess at most 1oz of useable cannabis.	A provider may possess 4 mature plants, 12 seedlings, and 1oz of useable cannabis for each registered cardholder who has named this person as the patient's provider.

Nevada

Possession & Use	Grow Limits
If the patient lives in a county with no dispensaries they will not be prosecuted for producing, possessing, or delivering 2.5oz of usable cannabis.	If the patient lives in a county with no dispensaries they will not be prosecuted for having less than 12 cannabis plants. The state of maturity of these plants is not limited. Patients are prohibited from growing and producing cannabis if a dispensary opens in the county they reside in.

New Hampshire

Possession & Use	Grow Limits
Patients are not allowed to purchase more than 2oz of cannabis during a ten day period. They are also not permitted possess more than 2oz of cannabis at any time. If the patient has a caregiver, there may not be more than 2oz of cannabis between the two parties at any given time.	Home growing is not allowed.

New Jersey

Possession & Use	Grow Limits
A physician will write instructions for a patient or caregiver to follow regarding the total amount of cannabis that will be dispensed. This will be measured in weight and be designated for a 30 day period. It will not amount to more than 2oz. If no amount is stated, the maximum 2oz amount will be dispensed at one time.	The patient or the primary caregiver are not permitted to grow or possess a cannabis plant.

New Mexico

Possession & Use	Grow Limits
Patients are allowed to possess at most 8oz of useable cannabis. Caregivers may transport at most 8oz for each patient listed on the caregiver ID card. If there is an exception to the 8oz limit, it will be noted on the back of the patient ID card.	Patients may apply for a license to grow their own cannabis. The license must be posted and kept near the growing area. A Personal Production License allows patients to grow up to 4 mature plants and 12 seedlings at any given time.

New York

Possession & Use	Grow Limits
Patients will not have a supply of cannabis that exceeds the dosage for a 30 day period determined by a physician. In the last 7 days of any 30 day period, the patient can possess up to that amount for the next thirty day period.	Home growing is not allowed.

North Dakota

Possession & Use	Grow Limits
Patients may possess up to 3oz of useable cannabis.	If patients live more than 40 miles from a dispensary, they are allowed to cultivate up to 8 plants.

Ohio

Possession & Use	Grow Limits
Patients are allowed to purchase cannabis topicals, tinctures, edibles, oils, and flowers. The smoking of cannabis is not permitted.	Home growing is not allowed.

Oregon

Possession & Use	Grow Limits
Adults (age 21+) may possess at the most 8oz of useable cannabis. A patient may possess at most 24oz of useable cannabis.	Adults (age 21+) who recreationally consume cannabis may possess up to 4 plants at one residence. A patient may possess at most 6 mature plants. These plants must be grown at a registered grow site address. Caregivers or Oregon Medical Marijuana Program Growers cannot be growing for more than 4 patients at one time and they also cannot grow more than 6 mature plants per patient.

Pennsylvania

Possession & Use	Grow Limits
Patients may have enough cannabis for a 30 day supply of the recommended dosage by a physician but only in the form of oils, tinctures, topicals, pills, and infused liquids.	Home growing is not allowed.

Rhode Island

Possession & Use	Grow Limits
A patient with a registry ID card will not be penalized for the medical use of cannabis as long as the patient is in possession of an amount that's less than 2.5oz of usable cannabis. A primary caregiver with and ID card will also not be penalized for assisting a patient they are connected to through a registration process. A caregiver also many not posses more than 2.5oz of useable cannabis.	Patients and caregivers may have no more than 12 mature plants. These plants must be stored in an indoor facility. Either patient or caregiver may possess a reasonable amount of unusable cannabis including up to 12 seedlings.

Maryland

Possession & Use	Grow Limits
Patients may possess a 30-day supply of cannabis for medical use.	Home growing is not permitted.

Vermont

Possession & Use	Grow Limits
A patient or caregiver may possess at most 2oz of useable cannabis.	A patient or caregiver designated to the patient may possess up to 2 mature plants and 7 immature plants. The total amount possessed between the patient and caregiver must not exceed the possession limit.

Washington

Possession & Use	Grow Limits
If a physician doesn't include a dosage recommendation when entering the patient or caregiver into the medicinal cannabis authorization database, they can obtain at a dispensary holding a medical cannabis endorsement a combination of: • 3oz of useable cannabis. • 21g of cannabis concentrate. • 48oz of cannabis sold in a solid form. • 216oz of cannabis-infused product in liquid form. Adults (age 21+) without a medicinal cannabis authorization may possess the following: • 1oz of useable cannabis. • 7g of cannabis concentrate. • 16oz of cannabis infused edibles in solid form. • 72oz of cannabis-infused product in liquid form.	Patients may grow in their home up to 6 plants for personal medicinal use and possess up to 8oz of useable cannabis produced from their plants. If a physician determines that the medical needs of the patient exceed those amounts, they must clearly specify that the patient be allowed to grow up to 15 plants and possess up to 16oz of usable cannabis in their home. Adults without a medicinal cannabis authorization are not permitted to grow cannabis plants for personal use.

Why Grow You Own Cannabis?

For medical cannabis users, there are three big reasons why you'll want to consider growing your own cannabis plants.

First off, it's a great way to have access to a **consistent** supply of cannabis. Not everyone has easy to access to cannabis even with a prescription though dispensary and co-op locations are popping up more frequently as laws allow for them. Still, if you're not close to one, it's convenient and empowering to be able to grow your own. You'll also never have to worry about running out once you get the hang of growing.

Secondly, there's the issue of **quality control**. When you grow your own cannabis, you're ensuring that you always have the type and quality of cannabis that you prefer to manage your health concerns.

The final reason to grow your own cannabis is **to save money**. Over time, it will be cheaper to grow you own cannabis especially if you have a long term prescription. Though the initial cost can be steep (upwards of $200 and averaging about $1900) since you're making many of the equipment costs up front, you're not spending hundreds of dollars all at once after that. To put it in perspective, at a dispensary prices for a single ounce of cannabis can range from $200-400.

An Overview of the Plant

Before you begin to grow your own cannabis plants, you'll need to have a basic understanding of the plant's anatomy and which strain is best to invest in for your needs. We'll outline these details next as well as the plant's growing cycle and what external factors it needs to thrive.

Seeds

Most growers purchase their seeds online. One of the biggest benefits to ordering your seeds online is that you can buy packets of all female plant seeds and you can research and pick whichever strain best fits your medical needs.

When looking at your seeds, you want to make sure they're healthy before you try to grow them. They should be hard and dark colored with little stripes. The stripes serve as a protective coating for the seeds and come off when rubbed. If your seeds are discolored- yellow or white- they're not going to germinate.

To keep your seeds healthy before planting, you can store them in the humidity controlled crisper drawer of your fridge. Make sure they aren't exposed to too much light though. Properly stored seeds can last about five years.

Cloning

You also might want to consider buying cloned plants from a medical cannabis dispensary or cloning your own plant. Cloning guarantees that you get exactly the strain of cannabis you want. Cloned seeds are the exact copy of the mother plant, so you can expect the cloned plant to grow similarly to the mother plant and to have the same physical and mental effects.

Anatomy of the Plant

There are four parts of the plant a grower should be able to identify and understand the purpose of other than the leaves, stem, and roots. These are the cola, calyx, trichome, and pistil.

Cola
The cola is a section of cannabis flowers at the end of a branch. The main cola forms at the top of the plant and is where the largest collection of female flowers bloom. Smaller colas bloom lower down on the plant, but there is generally one main one at the top. The number and size of colas are directly related to genetics and growing techniques.

Calyx

The calyx is a collection of small leaves that form in a spiral where the flower meets the stem. It is part of the bud and also part of the cola. To put another way, the calyx holds the buds together. Calyxes contain high concentrations of the glands that secrete THC. These glands are called trichomes.

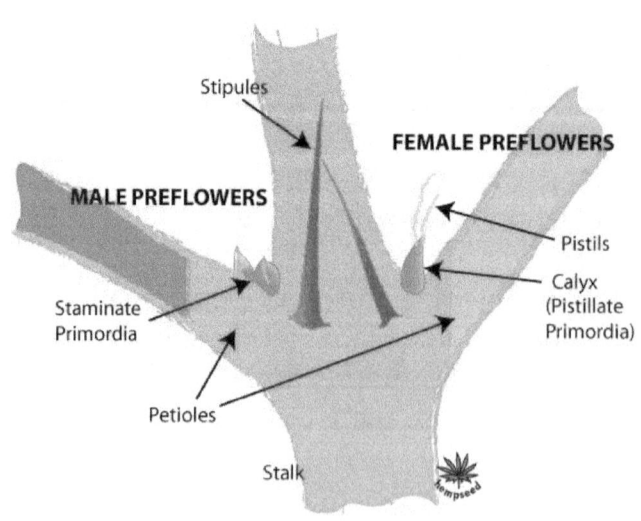

Trichome

The crystal resin you'll notice on cannabis buds is secreted through trichomes. Trichomes are mushroom shaped, translucent glands that originally developed to protect the plant from outside threats. Today, the oils they secrete- TCH and CBD- make medicinal cannabis as useful as it is in offering physical and mental relief to users.

Pistil

You'll be able to identify the pistils in the tangle that is the cannabis plant because of their color. The pistils are small hairs that grow out of the calyxes. Over the course of the growth cycle, the pistils change color from white to yellow to orange, red, and eventually brown. They're important for the plant's reproduction but don't have much of an impact on taste or medicinal effects of the cannabis.

Sex of the Plant

Whether smoked or ingested, cannabis comes from the budding flowers of female plants after they have been dried and cured. Male plants will not yield flowers and cannot be used for medicinal cannabis. Until the plants begin to flower and white pistil hairs grow or pollen sacs develop, you won't be able to tell if the plant is male or female. This should take about six weeks, when the vegetative stage starts to transition into a pre-flowering stage. As soon as you can tell the sex of the plants, remove the males so they don't pollinate the females. If a female plant gets pollinated, it will not yield as many buds or buds of the same intensity.

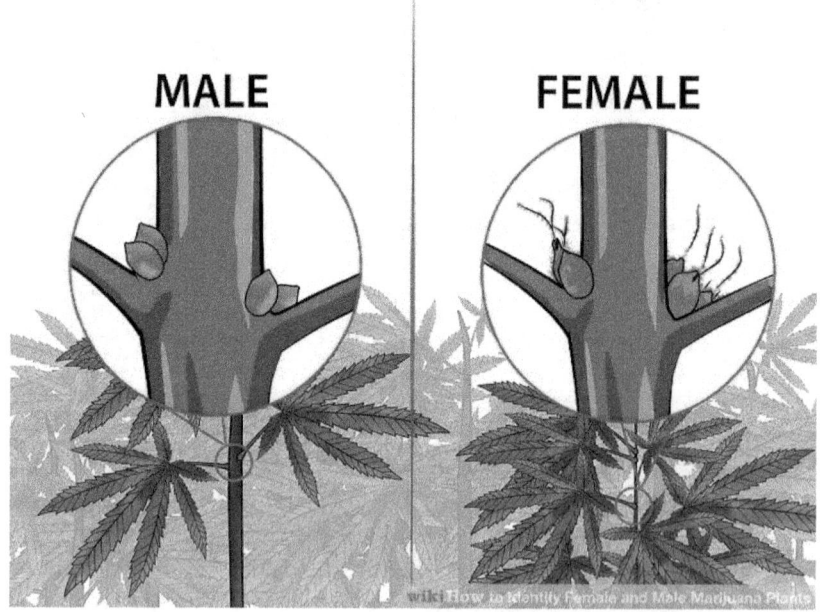

MALE   FEMALE

Four Strains

There are four main strains of cannabis: Indica, Sativa, Ruderalis, and Hybrid. Let's break down what these types mean for the plant's effects and characteristics.

Indica
Indica strain plants grow short and stout, they look more like a bush and feature rounded leaves. They are often ready to harvest faster, after about 8 or 9 weeks. Because of their size, shape, and light requirements, they are more convenient for indoor growing. The effects of Indica are mostly considered to be relaxing and bodily, for example a numbing effect.

Benefits of Indica-Dominant Strains
- Reduces seizures
- Relaxes muscles and body pain
- Stimulates appetite
- Relieves migraine headaches
- Relieves anxiety and stress

Sativa
Plants of the Sativa strain feature long, thin leaves. They also grow taller, 10-12ft, and need more light than Indica plants. Their maturation period is about 12-14 weeks. All of these features make them easier to grow outside. The effects of Sativa mostly impact a person's head and perception. It has a more energizing and appetite deadening effect on the user which can also translate into enhanced productivity or conversely, elevated anxiety.

Benefits of Sativa-Dominant Strains
- Encourages a perception of well-being
- Energizing
- Increases mental focus

Ruderalis (Auto-Flowering)
The strains of cannabis we've covered so far are labeled "photoperiod" strains. This means that they don't start flowering until they get tipped off by decreasing amounts of sunlight that winter is coming. Ruderalis plants don't need a change in light to begin flowering, they just begin flowering when they reach 3-4 weeks of age regardless of light schedule. Out in the wild, this means that wild Ruderalis plants were small, produced tiny amounts of bud, and contained low levels of THC. Modern cannabis breeders cross Ruderalis plants with various photoperiod plants to create a better balance between potency and speed.

Hybrid
Hybrid strains try to take the best traits of Ruderalis, Indica and Sativa to make for easier, hearty growing and balanced mind/body effects. Each hybrid plant is going to be a little different from the next and will lean more heavily towards one strain or the other. Depending on your condition and needs, there are even hybrid strains that have specific acting times. So if you need gradual relief over time or something more immediate and intense, you can choose a strain designed to work with time and your body in a specific way.

Environment Requirements

Cannabis plants require some specific elements to grow well. We'll talk about different grow mediums, how to track and adjust the water's pH, which nutrients to supply, how much light is required, and how to regulate temperature in our next section.

Grow Medium

Grow medium refers to what your plants will take root and grow in. You're probably thinking of soil, like most plants use, but you can also use soilless, hydroponics, aquaponics, and aeroponics to grow your cannabis plants. Here's a brief overview of each grow medium:

Soil
Soil is the most obvious choice and probably the most common and cheapest method. There is one trick to it though that can trip up new growers however: **don't get soils with slow releasing nutrients**. Slow release nutrient soils will negatively impact the flowering stage. Instead use non-nonsense high quality potting soil and standard store fertilizers. The downside? You might not get as great of a yield as you could with the following, more work intensive methods.

Soilless or Inert Medium

Soilless growing uses a medium like coco coir to encourage faster growth and larger yields. Coco coir is made out of coconut husks. You water the plants just as you would when using soil. Coco coir is a better growing medium because it is more effective when it comes to holding moisture, nutrients, and oxygen. It's still fairly inexpensive and is a generally cleaner way of growing. It also does not attract bugs. Unfortunately, the fertilizers you need with soilless mediums are more expensive and coco coir might be difficult to find in stores depending on where you live.

Hydroponics

Growing through a hydroponics system means there is no soil or vegetable matter involved at all. You're going to be using something like gravel in a pot or trough and adding a mix of fertilizers and water. If you use a pot, you pour water through the gravel and if you use a trough, you flood the trough.

Looking at the roots of a plant grown in hydroponics versus a plant grown in soil, you'll see that the roots in soil have one major tap root and the roots in a hydroponic system features many small, hairlike roots.

The advantages? Because of the root's exposure to more oxygen, the plant will grow faster. You'll have the biggest yield discussed so far using this system and and even less cleanup because there is no soil. You will also have a shorter vegetative state which means you can harvest sooner. A hydroponic system is more expensive however, and you'll need to invest in more hardware (pumps, troughs, and reservoirs) to grow effectively. With this system, you'll also need to monitor the pH and nutrient levels more diligently.

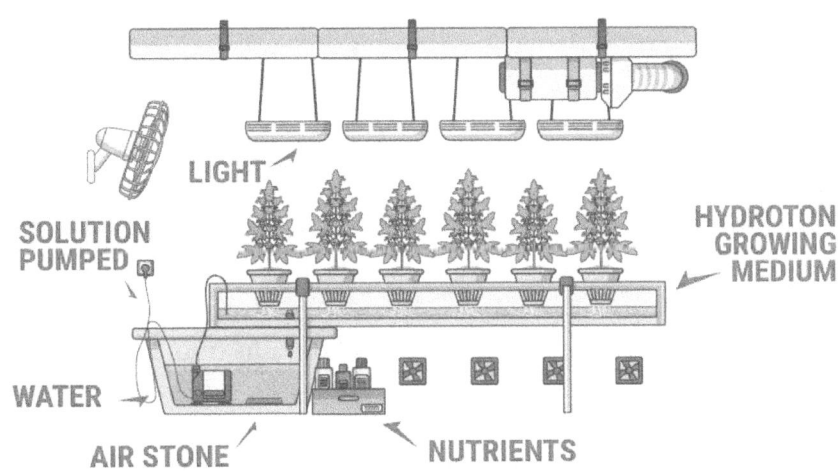

Aquaponics

Aquaponics is based on a symbiotic relationship between fish and the cannabis plants. The fish provide the plants with a food source- the water and their waste instead of soil- and the plant filters the water the fish live in. The naturally occurring bacteria convert the fish waste, uneaten food, and decaying plant matter into ammonia and other compounds for the plants to consume. Aquaponics best works in a climate controlled greenhouse and because of this, this system can work year-round.

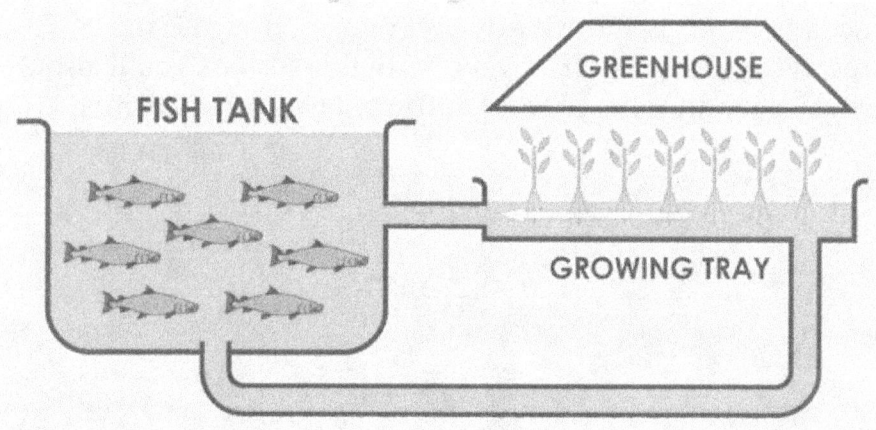

Aeroponics

In an aeroponics system, the plants are held in a mesh basket hung out in the air. The roots are watered and fertilized by spraying the roots. This speeds up the growth process by producing more roots than even a hydroponic system. This leads to amazing results! However, it requires more labor and equipment to be successful. This system also leaves the plants more sensitive to the nutrient and pH balance making it overall harder to care for.

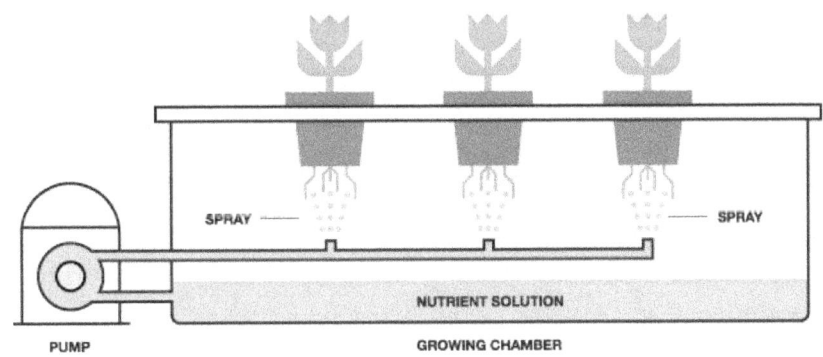

SPRAY

SPRAY

NUTRIENT SOLUTION

PUMP

GROWING CHAMBER

Tracking and Adjusting the pH

We've talked a little about the pH of the water, but it's time to explain it more in depth. pH is a measurement of the amount of free hydrogen and hydroxyl ions in water. If there's more hydrogen ions, the water is more acidic. If it has more hydroxyl ions, it's more basic.

How do you do test the pH? You can figure out the pH of your water you by picking up a basic pH tester kit and adding different solutions, called a pH Up and pH Down, to nudge the pH to where you want it.

Hard vs Soft Water

The amount of pH Up or Down you add to your water will vary depending on the water you start with. Water quality is generally considered "hard" or "soft". Soft means that there's not anything in the water. It's been treated to the point that the only ion present is sodium. Hard water contains more minerals like chalk, calcium, lime, and magnesium. If you're using soft water, you only need a small amount of fluid to adjust the water's pH because there are no other minerals to buffer the pH. If you're using hard water, you'll need to add more pH to account for the additional materials.

pH Up

pH up is much weaker than pH Down so you will need more of it to alter the overall pH. As a guideline, 2-4mL per gallon of water will raise the pH by 1 point.

pH Down

1mL per gallon of water will generally reduce your overall pH by 1 point.

Your initial adjustments will involve a lot of "guessing and checking". You'll add a little and then check the pH and adjust from there. It's helpful to make notes on the total pH Up and Down you've added to speed up the process later. You want to keep a root pH in the range of 6.0 to 7.0 if you're planting in soil. If you're planting it in hydroponics between 5.5 - 6.5 is the goal. There's not one number you need to keep it at. A range is actually good as different nutrients are absorbed better at different pH's.

More Tips for pH Management

- Gently shaking the water evens out the ratio of nutrients and the pH Up or Down. The extra oxygen is also good for the plant's roots.
- Use tap water or mineral water to create that extra buffer in the water. This helps to keep the pH relatively stable and easier to manipulate.

Nutrients

It's important that whatever the growth medium you choose, you enrich it with nutrients. For soil, this could mean composting your own or purchasing nutrients separately. If you're growing with hydroponics, you'll need nutrients specially formulated for hydroponic growing.

Regardless of medium, follow the nutrient feeding chart that will come with purchased nutrients to ensure the health of your plants. Tweak the chart as your plants develop to find the best results.

More Tips for Nutrient Management

- Don't over-do it with liquid nutrients. You can always add more but it's much more difficult and wasteful to remove nutrients. Start with half the recommended dose and only raise the amount of nutrients if the plant needs it.
- Add nutrients straight to your water. Don't mix them together as they can interact with each other and make the nutrients less useable to your plants.

Light

Cannabis is a fairly hearty plant but it does have specific sunlight requirements that take some planning and effort to accommodate. Cannabis needs 8+ hours of direct sunlight and they need to follow the sun's cycle from spring to autumn to grow and bloom properly. When the daylight shortens, this actually triggers the plants to start making buds in preparation for winter.

If you live in a sunny environment and can grow your plants outside all you really need to make sure you do in terms of lighting, is to plant in spring and harvest in the fall.

If you need to grow inside you're going to have to look at different lighting options and consider investing in a timer to simulate the hours of sunlight as it correlates to the seasons. There are three types of indoor grow lights: HID, Fluorescent, and LED.

High Intensity Discharge (HID)
HID grow lights are the most popular grow lights and offer a consistently great yield. **Metal Halide (MH)** creates a blue light that's great for encouraging growth in the vegetative stage while **High Pressure Sodium's (HPS)** yellow light encourages more growth in the flowering stage. **Light Emitting Ceramic (LEC)** creates a natural white light that can be used throughout the grow cycle -from seed to harvest- and produces UV rays. It takes the best of the previous two bulbs and creates an efficient hybrid.

Sample Setup
250W HID yields ~1-2oz per month
400W HID yields ~1.5-3.5oz per month
600W HID yields ~2.5-5oz per month

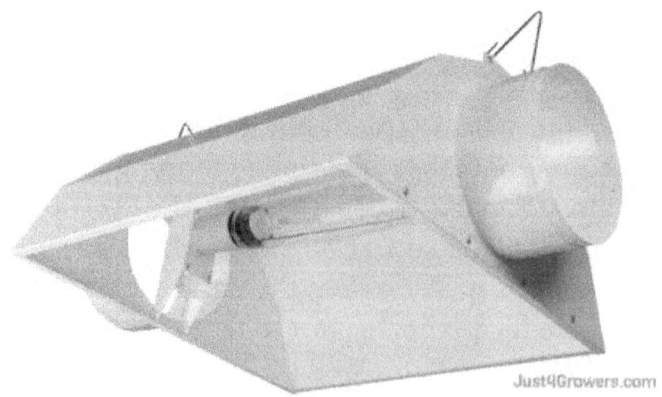

Just4Growers.com

Fluorescents

Fluorescent lights are great for small spaces and to get plants started. They also don't get very hot. They can be used until the plants are about 24inch tall, and at a distance of 1-4inch. Another big benefit of using fluorescents is the low impact they make on your electricity bill. There are two types:

- T5
- CFL

T5 lights are mostly used for small plants and seedlings and causes them to grow short and wide. In the flowering stage, the T5 lights must be kept close to the buds to produce a decent yield. Look for High Output (HO) bulbs to get the most lumens for your bulb.

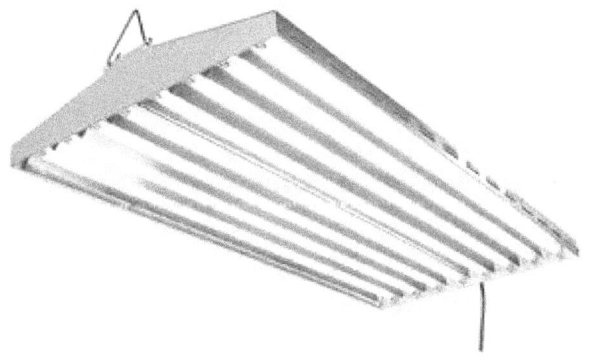

Sample Setup

For vegetative growth use bulbs that are labeled "Cool" or "Cool White" and 6500K.
For flowering growth use bulbs that are "Warm White" or "Soft White" and either 2500K or 3000K.

CFLs can also be kept close to the plants allowing you to cut back on the total amount of space you'll need. These lights can conveniently be used from seed to harvest but won't be able to produce the biggest yields.

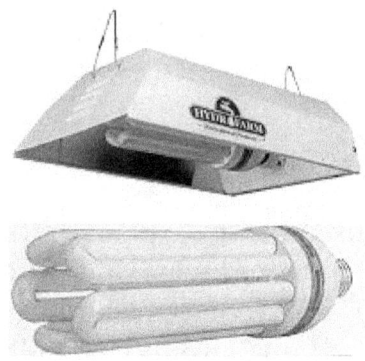

Sample Setup
2 x 40W Daylight CFLs
2 x 42W Soft White CFLs
Equals ~1oz of bud for each 150W of CFL

Light Emitting Diode (LED)
LED lights are the latest in lighting technology and are constantly being developed. They feature excellent yields for the same amount of electricity as CFLs, a customizable spectrum of light, and they don't emit too much heat so you don't need to invest in an exhaust system. In fact, many LEDs come with a cooling option like built in fans to push heat up and away from plants to help create and maintain a stable temperature.

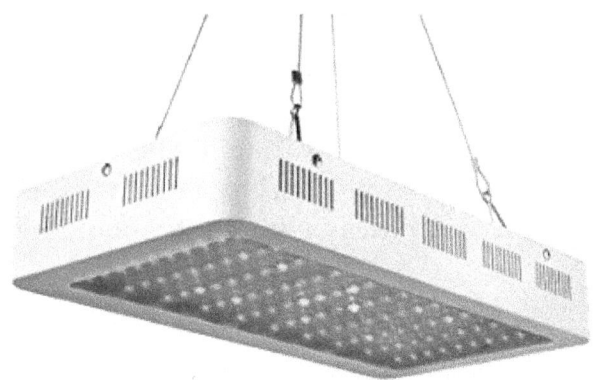

As great as their yields can be, it's been difficult for growers to find the right light spectrum, size of the diode, and angle configuration for each model to grow effectively because they can be so different. It's also important to note that LEDs must be kept at a distance of about 12"-18" from the tops of plants to avoid bleaching and burning. Until the kinks are worked out and the process is streamlined, there are a lot of unknown risks as much as there are celebrated perks to using LEDs.

You can also consider a combination of LEDs and HPS bulbs to improve the overall quality of your buds. LEDs are commonly thought to increase the potency of your strain while HPS bulbs produce better looking buds.

Sample Setup
2 x 125W LEDs ~0.5g per watt

Distance Between Plant and Grow Light

Because there can be so much variability between plant and light distance, we've compiled a cheat sheet to give you some baseline idea of how far away your lights should be from your plants.

HID Grow Light	Distance
150W	8-12"
250W	10- 14"
400W	12- 19"
600W	14- 25"
1000W	16"- 31"

LED Grow Light	Distance
1W Bulbs	at least 12"
3W or 5W Bulbs	at least 18"
High Wattage (300W+)	up to 30"

Fluorescent Grow Light "Hand Test"
With fluorescent grow lights, you can keep the light as close as you can get it as long as it passes the "hand test". Put your hand where your plants are for 30 seconds and if it doesn't feel too hot for you, the light it just fine for the plants. If it is too hot for you, back the light up and try again.

Temperature and Ventilation

Similar to how the amount of sunlight triggers growth in the cannabis plant, temperature that can fluctuate will serve the plant better at different stages of growth as well. When the plant is young, before flowering, it's best to keep them in temperatures in the 70-85°F range. In the flowering stage, you want to decrease the temperature a bit, somewhere in the range of 65-80°F.

Ventilation is more important when growing large amounts of cannabis because you need a lot of CO_2 coming in and you need to allow the hot air to escape. There are three things you will want to invest in when considering ventilation: a fan, filter, and duct.

Fan
Depending on how much cannabis you are growing indoors, an industrial intake/outtake fan might be worth investing in. For smaller supplies, leaving a door open twice a day will be sufficient. Having an oscillating 4-6in fan blowing will help control the temperature and humidity adequately for most indoor growers too. To reduce noise and save on electricity costs, a fan speed controller is also an option. While the plants are young or during their dark periods, you can set the fan to lower speeds or off to save some money.

Filter
The smell of cannabis is very strong and distinct, especially late in the grow cycle. Not everyone appreciates this and installing a filter to the front of a fan will help to remove the smell from the air. Keep in mind, you will have to replace the charcoal in the filters from time to time.

Duct
Ducts are used to lead air out of windows which will also help to control the temperature and smell in your indoor grow area.

Indoor vs Outdoor Growing

Now that we've covered the basics of the plant and it's needs, it's time to figure out where you'll be growing your plants: indoors or outdoors.

Indoors

Growing indoors is a great option if you're tight on space, don't have access to fertile land, or don't live in a particularly sunny environment. To get started you'll need a space like a spare room, closet, garage, attic, garden shed, or bathroom. Whichever space you choose, you must have a window or a vent where you can get rid of used air and electricity so you can plug in your lights. Alternatively, you can buy or make a space specifically designed to grow your plants in, like a grow tent or grow bucket.

Grow Tent or Grow Bucket
A grow tent is a lot like what it sounds like. It's lightproof, has reflective walls, a waterproof floor, and there are places to hang lights and vent your tent. A grow tent also features an opening at the bottom for fresh air to enter and an opening at the top for used air to exit. There are also openings for electrical cables to feed through.

A grow bucket or space bucket is a planting tool you can make or purchase that consolidates your grow medium, plant, and light into one space efficient bucket. You can shop for premade versions of either of these row spaces or build your own.

Tips for Transplanting

Transplanting is an important step in your plant's life because it encourages root growth. Generally, make sure the containers you choose to grow your plants equals about 1 gallon per 12in of height. Different strains of cannabis plants grow differently, but this is considered a good rule of thumb. Roots grow downwards and if not given enough room, they can get root bound in a container that is too shallow. You can start growing in something as small as a solo cup but eventually you will graduate to 5in pots, 1 gallon pots, and even as big as a 5 gallon pot depending on the plant and how you choose to grow it.

Whatever container you choose to grow your cannabis in you have holes drilled close to the bottom so water can drain out. That said, also make sure you have a tray underneath so that the water that seeps out doesn't leak onto the floor.

When thinking about the actual soil your plant will be transplanted to, make sure it has been soaked with water and balanced at a pH of 6.0 for at least one hour to prevent the roots from drying out.

Indoor Lighting

Another challenge of growing indoors is light. You'll need to invest in proper lighting to encourage your plants to grow. With growing indoors, you again have that great sense of quality control but on the flip side, it's more responsibility for you. Make sure you get a timer and take safety precautions like purchasing a surge protector.

24-hour adjustable timer

Outdoors

To grow outdoors, you'll need a very specific environment. Not only does that mean space and good soil, but you need to live in an area where you get at least eight hours of direct sunlight. Growing outdoors also means you're limited to growing with seasons: planting in spring and harvesting in fall as opposed to growing year-round with artificial lights and timers.

This is the cheaper option if you can swing it but you also run different risks than if you grow indoors. There's the issue of privacy from neighbors and people potentially stealing your plants as well as the environmental risks of animals, bugs, and bad weather. All of these factors can have adverse effects on your plant's water intake and temperature exposure.

It's best to start planting your seeds in the spring. In Australia and South Africa for example, Spring is in October. In Germany and The USA it is late March.

Cost

Generally, it costs less to grow outdoors. You're not paying for electricity or to moderate the humidity- just soil, seeds, and nutrients. Financially, you'll be somewhere in the range of ~$200.

Growing inside will cost you more but it also depends on how ambitious you are, how much you want to grow, and which lighting and grow medium you've chosen. DIY is a strong value in the world of homegrown cannabis, there's a lot of room for making your own anything. You can buy a space bucket just as well as make your own, clone your plants instead of continually buying seeds, and make your own compost for a few examples. So if you're a medicinal cannabis user or caregiver on the East Coast or in the Pacific Northwest where the weather isn't great but you also don't want to spend thousands of dollars, you might want to consider taking DIY a step further. Instead of limited yourself to DIY growing, DIY as much of your grow setup as you can too.

From Growing to Harvesting

The turnaround time for growing your own cannabis is generally around 4 months. If you're working with clones, it can take somewhere between 2 and 2 1/2 months. We'll break down the stages from growing to harvesting next.

#1 Germination

Germination is the process of getting your seeds to sprout, which can take between 12 hours and a few days. The seeds need three things to properly germinate: warmth, darkness, and moisture. When done properly, you'll see a little white tendril pop out of the seed, that's the tap root. Next we'll go over three ways to get your seeds to germinate: a seeding plug or starter cube, planting directing into the soil, or soaking overnight.

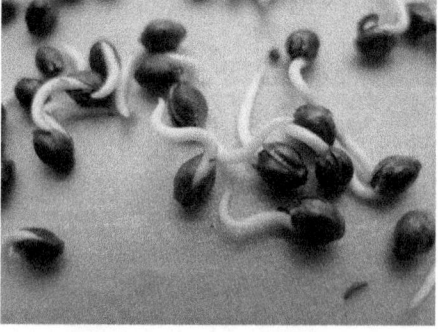

Before you begin, make sure your hands and workspace are clean. Especially if you're a smoker, nicotine can damage cannabis seeds and plants so you'll want to be clean of that.

wUsing a Seedling Plug or Starter Cube

To start them in a seeding plug or starter cube, simply place the seed in the plug or cube and water as directed. There's even a hole for you to insert the seed and get things going.

Planting in Soil or Soilless Mediums

For soil or soilless mediums, you can also plant directly into the final medium. Just plant the seeds knuckle deep in moist soil. The benefit of planting right into your final medium is that you don't have to worry about transplanting and any issues that could arise from stressing the plant.

Soaking Overnight

Finally, you can soak overnight. Using a glass cup and slightly warm filtered water, place your seeds into the water and let them sit overnight. The most viable seeds will float to the top and then gradually sink down. After 24 hours, check in on the seeds to see if the tap root has broken through. Leaving the seeds in water for too long can cause them to drown, so if they haven't germinated after 24 hours place the seeds in a warm, moist spot to finish germinating.

Once the seeds have sprouted, plant them as soon as you can into a bigger container. Don't touch the tap root with your fingers as it can be damaged easily. Plant the seeds so that the root faces downward and is about knuckle deep in your growing medium. Whichever method of germination you choose, it does make sense to keep the new seedling in a small container at first to allow the roots to have access to the most oxygen.

It's in no way a death sentence to plant your seedlings in a bigger container, but they might grow slower in the beginning. Don't forget to poke holes in the bottom of your container to allow for the water to drain out

The seed has officially germinated and begun the seedling stage when the root has fixed itself into the soil and pushes two leaves through to the surface.

#2 Seedling Stage

The seedling stage features round leaves, called cotyledons. After these initial leaves, all others will start to have the characteristics of cannabis leaves. Specifically, the blades or "fingers" of the plant leaves which will increase as the leaves keep growing.

Sometimes the stems are very weak during this stage and can be supported by tying them with thread to a thin wooden stake. This stage can lasts for 1-3 weeks indoors and 4-6 weeks outdoors. At the end of the seedling stage there will be 4-8 new leaves.

At this time, the seedling requires moderate humidity levels, medium to high light intensity (18 hours), and adequate soil moisture. Lights in the blue spectrum that produce little heat like T5 fluorescents, are a great choice for this stage because they won't dry out the plant.

T5 lights can be kept 2in from the plant. You can also use HID lights though they will produce more heat and you will need to use the hand test to monitor its temperature. It's also a good idea to have a weak fan blowing on the seedlings to help them get fresh air and to have something to resist to will encourage strong root growth.

Water your seedlings with filtered water when the soil is dry and warm. If the soil is cool and a little damp, it doesn't need to be watered. You want the water to be filtered so chemicals like chlorine and flouride don't contaminate them. If your seedling's leaves are hanging down, they are definitely thirsty and need to be watered immediately. Make sure the containers they are in are draining properly as well. This stage of the growing cycle is the foundation for the rest of the plant's life.

The seedling stage has ended when the leaves have the maximum amount of fingers (the number varies from plant to plant), has a stem width of 4-6mm, and a height of 3-4 nodes. You will also notice that the roots will begin growing out of the holes in the container it's in which is another indicator that it's time to carefully begin the transplant process.

#3 Vegetative Stage

The vegetative stage is marked by leaf and stem growth, not bud growth. During this period of 1-2 months, your plant will get bigger and leafier so they can support the weight of the buds. Amazingly, a healthy cannabis plant can grow up to two inches in a day! Keep temperatures in the 70-85° range with a humidity of 40-70%. In terms of nutrients, make sure your levels of nitrogen are high, phosphorus is at a medium level, and potassium is high. A standard vegetative nutrient formula will follow this formula.

Your plants should be getting between 18 and 24 hours of light a day from a bulb on the blue spectrum during this period. Cannabis will stay in a vegetative state until the plant begins to experience nights that are shorter than 12 hours. It's important to keep this consistent. Any inconsistencies can stress out the plant and potentially turn it into a hermaphrodite, which will seriously alter the quality and quantity of the buds. Indoor growers can control how long the vegetative stage is whereas outdoor growers are subject to nature.

If you're growing indoors, it's best to force your plants to flower when the tips of the leaves are touching each other. They will continue to grow in the first 2-3 weeks of the flowering stage anyway but you don't want your plants to outgrow your room. In terms of nutrients, make sure your plant is getting plenty of nitrogen.

Cloning

If you want to clone a plant yourself, the vegetative stage is when to do it. Using a clean razor blade, cut a branch off of the stem during the vegetative stage. Cutting a branch during the flowering stage will cause stress to the plant. You want to cut down along the natural angle where the branch meets the stem, this should be at about 45°. Then cut down the middle of this branch, just a little before the end of it, before planting. Cutting the branch twice like this increases surface area and encourages root growth. When transplanting it to a smaller pot or tray and grow medium, you can also use rooting hormones or cloning gels to encourage root growth.

Clones take about three weeks to develop healthy roots. Once the roots are established, it's time to transplant them to their next container. Here they must be under light for 18 hours a day until reaching a height and width that just touches the other plants. Once it reaches this stage, it's ready to move on to flowering just like a plant you had from a seed.

#4 Pre- Flowering Stage

If you're growing outdoors, you can expect the flowering stage to begin as the days get shorter and we head into fall. If you're growing indoors, this is the time to change your light schedule to 12 hours on, 12 hours off so the plant begins to flower. Avoid any light interfering with the actual or artificial "night" that needs to occur during. That means no street lights or spotlights coming off of a back porch or something similar. If the plant's nights are encroached by light they might revert back to the vegetative state and delay flowering.

During the first few days to two weeks after switching to a 12 on, 12 off light schedule, the plant will rapidly grow in height and branches as the plant grows. This period is commonly called "the stretch". Keep your plants spread out and gently bend any stems down and away from the center of the plant that look like they are "reaching" or stretching very high.

By creating a flat canopy of leaves you will increase your bud yield by as much as 40% because it evens out the light distribution. Keep in mind that you won't be able to make adjustments like this later when the stems harden and become woody.

The sex of your plant will also be revealed as it's getting ready to share its genes. Female plants will begin to grow white pistil hairs. Male plants will grow pollen sacs. If you are growing both, make sure to remove the male plants to avoid pollination.

Female Male

#5 Flowering Stage

You can expect the duration of the flowering stage to last between 6-10 weeks. Drop the temperature down between 65-80° and the humidity to 40-50%. As far as nutrients go, keep nitrogen levels low, phosphorus in a medium to high range, and potassium high. A standard bloom nutrient formula will follow this recommendation. We'll explore what happens in this time period next.

Weeks 3-4 Beginning to Bud
The rapid growth upwards you've been seeing will begin to slow down after 3 or 4 weeks and buds will start to emerge. At this stage, your plants may also start to smell. During this time, you'll need to pay more attention to your plant as it's environmental and nutrient needs will change. Do not change your nutrient solution until the plant shows very clear signs of flowering. Check out our troubleshooting section later on if you notice negative changes like yellowing.

Weeks 4-6 Buds Fatten

Now you can expect the buds to become more substantial and all of the white pistils will be sticking up straight. Your plant will no longer be making new leaves or stems so make sure you are watching them carefully for any problems during this time. You may also prune away all the buds that are in shadows or leaves that are dead or otherwise compromised.

Weeks 6-8 Buds Ripen and Pistils Darken

With all of the plant's attention turned from general growth to developing buds, it's more susceptible to pH and nutrient problems. Watch out for yellowing leaves which can be a sign of light burn or a nutrient deficiency. In the final weeks of flowering you will also want to drop the humidity to less than 40%.

On female plants you can expect to see a lot of sticky resin secreting from the trichomes on the outside of the leaves and buds. This resin is THC, the ingredient that medicinal cannabis users seek for it's therapeutic qualities. The potency of your plant is determined by the amount of time it's spent flowering and if it's been fertilized, which is why it's important to remove male plants.

After 8 Weeks

By now the buds are fat, the trichomes and the pistils are mature, and your plant should have a very strong smell. Some of the leaves might yellow during this time but as long as the yellowing isn't impacting the buds, you're doing fine. Don't raise the nutrient level to try to combat minor yellowing.

It's a common practice among growers to stop providing nutrients to their cannabis for the last few days or up to two weeks of flowering before harvesting in order to let the plant flush out nutrient build up. If there's a nutrient build up it can create a chemical taste in the final product. Don't flush the plant of its nutrients until the pistils are nearly all darkened and curled.

#6 Harvesting

When to harvest is highly dependent on the strain you've chosen and your personal preference in the cannabis' effect. You can start as early as after two months of flowering or wait longer than four months after initial flowering. The timing of the harvest determines the taste, smell, weight, and effects of your cannabis and should be considered carefully.

The best way to decide when to harvest is to look at the pistils. These little hairs are white when they first appear, but over time they curl and change color. They can turn yellow, pink, red, purple, and brown. A guideline regardless of strain, is to harvest when 70 to 90% of the pistils have changed color, but here's the total breakdown:

Look at the Total Percentage of Color Change
- If 0-49% of the pistils are brown, they're not ready yet.
- When 50-70% of the pistils are brown they can be harvested but are considered young. At this age they will give you a light taste and a mellow high.
- When 70-90% of the pistils are brown they're ready to harvest. This is the peak of the plant's taste and effect.
- After 90% you're edging into it being too late to harvest. The taste is heavy and the effects are compromised. Don't wait any longer to harvest.

Other Tips
In these last two weeks of growing, some growers feed their plants a teaspoon of blackstrap molasses per gallon of water. The sugar helps the buds improve in size and flavor.

Trim the leaves from the buds using small scissors. The big leaves are have no active ingredients that would be beneficial to your medicinal cannabis. The smaller leaves that are covered with resin can also be cut away but can be used in smoked cannabis or for hashish.

#7 Drying and Curing

To begin drying, cut off the trimmed branches on the individual buds from the plant. In a well ventilated, cool, dark place hang these pieces upside down. Make sure there is space between each bud, they should not be touching.

The drying process should take between four to ten days. Check the stems of the buds. When the smaller stems are fully dried, they will snap when you bend them, the thicker ones shouldn't. Now you're ready to start actively curing the buds.

Curing is meant to enhance the quality of the buds before you smoke them. The curing process begins when you stop watering your plant, so while it is being dried, it has also begun to cure. To continue the curing process, you'll need jars and lids. Fill these jars to what looks like 80% full and tighten the lids. Store the jars in a cool, dark place.

For the next two weeks or so, open the jars once a day for a few seconds to get fresh air into the jars and to let any built up moisture evaporate. Similar to the drying process, you'll need to watch your jars and levels of humidity so mold doesn't begin to grow. If the buds feel wet at any point you can open the tops of the jars and allow the buds to be exposed to air until they're dry on the outside. After a week of the buds feeling dry to the touch, you can open the jars once a week instead of every day.

A properly dried and cured bud will not have the grassy smell any longer and will result in cannabis that does not have a harsh taste.

Troubleshooting

Cannabis plants are very resilient. If you are an attentive grower and fix problems as they arise, you will have healthy buds. Next are some tips and tricks to keeping your plants healthy and help them recover when problems do arise.

Seedling Issues

Transfer Timing
Do not transfer seedlings until after your plants have grown the initial set of leaves. Transplanting to early can cause stress and damage to the plant.

Nutrient Amounts
In terms of nutrients, you don't want to rush this either. If you're planting in a soilless medium or through hydroponics, only add nutrients at seedling strength, which is about ¼ of regular strength.

In the very beginning, before your seedling have many leaves, you may leave a CFL light 6in away from the plant as long as it passes the hand test. After its first two sets of leaves have developed, move the lights as close as 2in as long as it passes the hand test.

When in doubt, check the pH of the soil to troubleshoot any perceived seedling issues.

Watering

Root Rot

If you don't have healthy roots, you don't have a healthy plant. If your plant becomes root bound it means that the roots become tangled, matted, and are growing in circles because there was no room for them to spread out in the container they were in. Root bound plants that don't get their root untangled usually fail to overcome their choked condition.

You can tell if you plant is root bound by looking at the bottom of the pot. If roots are poking out of the bottom, this is a bad sign. Ease the plant out of it's pot to see what the roots look like. If you see more roots than soil, it's definitely root bound.

Soak the plant in a bucket for a few minutes before transplanting them and take off the bottom of the root ball. Then you have two options. You can try to massage the root ball with your hands to loosen the roots and open up the ball first. If there are big, long roots circling the bottom cut those short. Or you can get a knife and make long, vertical cuts down the sides of the rootball and replant. Realize that you will have to water these plants more than other plants because of the damaged root system.

How do you keep the plants healthy and avoid root rot? Proper drainage is a great place to start for potted plants. Make sure the water can drain freely from the potted plants and avoid over watering. Also make sure that the grow medium is loose. If it gets too compacted, there's no literal breathing room for the plant. You can add extra perlite if you're working with soil, in hydroponics make sure there are a lot of bubbles by using an air pump.

Under-watering
You might begin to suspect yours plants are being under-watered if you're seeing the following signs:
- Drooping plants
- Thin, papery leaves
- Yellow leaves
- Dry soil

Under-watering typically happens when folks leave town for a long time or when plants are outgrowing their pots. You can solve under-watering a few different ways: watering more often, giving more water and a time, setting a watering schedule and sticking to it, and transplanting your plant into a bigger pot.

Over-watering
You run the risk of overwatering if you water your plants too often or if it's unable to drain properly. Make sure you're waiting to water your plants when the top of the grow medium is dry and about a half in down is also dry. You also shouldn't have too much run-off if you're water in a pot, only about 10-20%. To prevent or stop over-watering due to poor drainage you can mix in perlite to loosen the soil and remove bark or wood chips if present. You also might want to try using a 'smart pot'. A smart pot is made out of a porous fabric which allows the medium to remain contained while water flows freely throughout.

Nutrients and pH

You need to feed your plant a combination of mineral and non-mineral nutrients. The minerals come from the soils or grow medium and the non-minerals come from the air and water. Here's how it breaks down:

Minerals:
- Calcium
- Magnesium
- Nitrogen
- Phosphorus
- Potassium

- Sulfur

Non-Minerals
- Carbon
- Hydrogen
- Oxygen

When you look at fertilizer bags and nutrient solution bottles, you'll see a percentage breakdown of how much of nitrogen, phosphorous, and potassium are in each product which will look something like 10-5-5 meaning 10%, 5%, 5%. They will be listed in the same order: nitrogen, phosphorus, potassium, and abbreviated as N-P-K.

For hydroponic growing: The nutrients you purchase will come in two parts, usually labeled A and B. This is because if they are mixed together at a high concentration, they won't work as intended. When mixed with the right amount of water, this won't happen. That said, the nutrients you use must therefore be water soluble. You'll find that there will be solutions labeled "Grow" and solutions labeled "Bloom". Grow solutions are high in nitrogen while Bloom solutions are high in phosphorus. Start your nutrients at about a quarter of the recommended level to avoid overfeeding.

For soil growing: Soil already contains organic materials that the plant can get it's nutrients from like compost and manure. Some of these substances must be processed by soil-dwelling microbes to actually be useful to the plant. Sticking to organic fertilizers and nutrients prevents these substances from building up faster than the microbes can them break down. You can either make your own our purchase nutrient solutions. Either way, these are not hard to find at a home and garden center. Here's how the nutrients break down if you want to do it yourself:

- Nitrogen- blood and fish meal.
- Phosphorus- bat guano and bone meal.
- Potassium- kelp meal and wood ash.
- Magnesium and Calcium- dolomite lime.
- Magnesium and Sulfur- epsom salt.

Having the correct pH directly impacts the plant's ability to absorb its nutrients. Depending on the grow medium you're using, your approach to nutrients and pH monitoring is going to vary.

Soil
If you suspect there's a problem with the nutrients or pH with plants that live in a potted container, you'll want to flush the system. Get two times the size of your container's pH nutrient water at half the normal strength and water the plant with this. The water should drain freely to the bottom of the container. If it is taking a long time to drain between waterings, you might actually have a problem with drainage.

Hydroponic
When in doubt, do a complete overhaul of the water reservoir. Replace the old water with fresh water featuring the proper pH and nutrients. To tell if a plant is recovering, you won't see damaged leaves get better but you'll see that the problem hasn't spread, new leaves will be coming in healthy. This is a common misconception that can confuse new growers.

Nitrogen Toxicity

If the leaves on your plants start to bend at the ends and look like a claw, if the leaves are a dark green, if the stems are weak, and if the overall growth is slow, you're likely looking at a case of nitrogen toxicity.

If left untreated, the leaves will turn yellow and the plant will die. It's a common mistake to give a plant too much nitrogen, especially in the flowering stage. This is where the danger of using a time release soil comes in. With a time release soil, the plant will continue to get high amounts of nitrogen throughout it's life instead of varying levels as it matures. Here's the nitrogen plan you should have your plants on:

Vegetative Stage: High levels are good here. Any complete plant food will work just fine.

Flowering Stage: Drop down to lower levels here, mixes labeled "Bloom" or Cactus nutrients.
You can't reverse nitrogen toxicity, but you can prevent new leaves from being affected.

Calcium Deficiency

Calcium deficiencies tend to show up on new leaves, so you want to check your new leaves for any:

- Dark green leaves with dead or brown spots
- Crinkling
- Curled tips
- Stunted growth
- Weak or hollow stems
- Underdeveloped, weak roots
- Slow or undeveloped buds
- Decay

Calcium helps the plant withstand stress and provides structure for the plant. You likely won't be overdoing it on the calcium if you are using normal amounts of nutrients or regular soil.

Calcium deficiency is likely to occur:

When growing outdoors: because the soil can be more acidic, which is a pH below 6.2.

When using soil: and you haven't added a calcium supplement like dolomite lime or the water is acidic. When growing in soil, calcium is best absorbed by the roots in the range of 6.2- 7.0 pH.

When using coco coir: and you haven't added a calcium supplement or the water is acidic.

When using a hydroponic system: and you've neglected to use nutrients that will supplement the calcium or growing in water that is acidic. When growing in a hydroponic system, calcium is best absorbed by the roots at a pH between 6.2- 6.5.

You'll notice that acidic water and soil are the most common denominators. This is because when the pH is off, the plant cannot absorb calcium through the roots. Therefore, a good first step in figuring out if your problem is calcium deficiency, you should check the pH levels. Your second step should be to flush the system with clean, pH'd water with normal levels of nutrients.

After you've taken action against calcium deficiency, pay attention to the newly growing leaves. The brown and dead spots already on the leaves will not go away, but new leaves should be coming in healthy and strong within a week.

Temperature and Humidity

Too much heat will force the plant to drink more water than normal and will cause the leaves to curl and turn brown. If exposed to too much heat or light, the leaves and buds can also bleach. Plants exposed to high temperatures for a prolonged period of time, may not recover from the stress.

In terms of humidity, you want to remember that high humidity helps plants grow faster when they are young but it can cause mold in the budding and flowering stage. For young cannabis plants, keep the humidity between 40 to 70%. Once the plant begins to flower, keep the humidity below 45%.

If you're growing indoors you'll want keep an eye on the temperature and humidity by investing in a temperature and humidity monitor. You'll vent out the hot air with a fan or exhaust system. A basic oscillating fan will be fine. You just need to get that hot air circulated and a fan that can move will prevent hot spots. A humidifier or dehumidifier will also help to manage the humidity levels in the air. A simple room fan or a computer fan for smaller installations will be fine. Generally the bigger the fan the lower the noise if you move the same amount of air. For example: if you use a 80mm computer fan it needs to turn faster than a 120mm fan to move the same amount of air, thus creating more noise. You can regulate the speed of a computer fan by using potentiometers.

A simple potentiometer

If you're growing outdoors, you have a lot less control over the environment and weather but you can work with the forecasts to prepare your plants. For example, if it's going to be hot and dry water your plants early in the morning or late in the evening. This way you're minimizing the amount of water that will be evaporated during the hottest hours of the day. You can also hang up a sheet to further shade your plants from the sun's beating rays. With potted plants, you can just move them into the shade. In humid conditions, do the best you can to make sure nothing is blocking your plants from the wind and avoid over-watering. You'll also want to make sure that the leaves of the plants aren't touching each other as this can cause mold to flourish.

Light

One of the most common mistakes new growers make is not giving their plants enough light. If your plants seem to be slow growing or not taking in a lot of water, they probably just need more light.

You can give your plants too much light however and this is just as damaging as too little light. Dry, cupped or turned up edges, and brown leaves are all signs your plant has been exposed to too much light. Especially with young plants, lights can be too close and too intense and burn the plants. This is mostly an issue for indoor growers but if you're transferring plants from pots to the ground, there is some risk of stress occurring during this transition depending on how the plants were lit before transitioning to plots in the ground.

Bugs

Bugs are a common pest to all plants, cannabis plants included. As soon as you think you have a bug problem, identify the type so you can quickly determine the best course of action. You can also use SM-90. SM-90 is somewhat of a cure-all in the growing world when it comes to mold and bugs. Mixing one part SM-90 and five parts water, all you need to do turn the lights off and mist it on the leaves of the plant. SM-90 is safe for humans and animals but it does a great job of killing bugs and mold on contact and preventing further attacks.

Flowering for Too Long

Unfortunately, cannabis can over-mature. When this happens, the medicinal value of the plant diminishes. If you think this is happening, look to see if the plant has stopped producing calyces (calyx), if they look stretched or if the stem swells and leaves yellow and fall off.

Plants are Falling Over

If your cannabis plants are so heavy they begin to fall over at the end of the flowering stage, you can get specially designed hooks to hang from the ceiling- called "plant yo-yos"- to hook around your buds and support them.

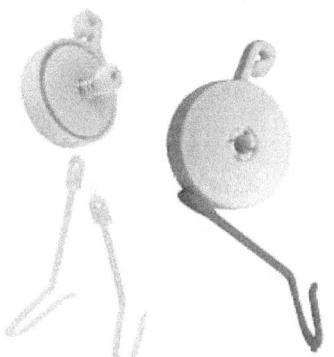

Some Plant Yo-Yos

Recommended Retailers

Seeds
- Seedsman (www.seedsman.com)
- MSNL Seeds (www.marijuana-seeds.nl)
- Nirvana (www.nirvanashop.com/en)
- Sensible Seeds (www.sensibleseeds.com)

Starter Cubes for Seed Germination
- Rapid Rooters
- Jiffy Pellets (Coco Coir or Soil)

Soils
- Happy Frog Potting Soil (for young seedlings)
- Fox Farms Ocean Forest (for higher levels of nutrients)

Nutrients
- Botanicare (Soil and Hydroponics)
- Canna Coco (Coco Coir)
- Fox Farms Nutrient Trio (Soil and Coco Coir)
- General Hydroponics Flora Trio (Coco Coir & Hydroponics)
- General Organics Go Box (Soil)
- House & Garden Nutrients (Soil, Coco Coir and Hydroponics)

pH Testing Kit
- General Hydroponics pH Testing Kit
- Hanna Instruments HI99104 Educational pH Tester
- Sunleaves pH Test Kit

HID Light Brands
- iPower
- Growlite
- MaxLume
- Plantmax
- Solistek
- SunMaster Grow Lamps
- SunPulse Lamps
- Ushio

Fluorescent Light Brands
- Agrobrite
- DuroLux
- Hydro Crunch

LED Light Brands
- Advanced Platinum LED Grow Lights
- GalaxyHydro LED Grow Lights
- Kind LED Grow Light

Let's get cooking!

What is hash?

To fully understand what hash or "hashish" is, it is essential to know the dualism that exists with the cannabis plant. First is the plant's physical structure, which is a rich, leafy and fibrous material that contains essential amino acids and plenty of benefits. Second is the essence of the plant, called trichomes, which contain most of the cannabinoids that gets everyone buzzed.

The trichome glands are responsible for producing the aromatic terpenes, mood-enhancing THC, and medicinal CBD that makes the plant an excellent therapeutic agent. To make hash, those glands (also known as kief) are separated from the plant matter and condensed. Hash is more powerful by weight than the cannabis flowers.

The traditional way to make hash is to rub the cannabis buds by hand until the oils and kief are thick enough to chafe off. The kief and oils are then gathered and pressed into balls. This method has been used since thousands of years ago in India and is still common in many parts of the world.

Another method of making hash is the "dry-sieve hash." The dried cannabis plants are beat over screens or large sieves, dropping the kief to the collection trays placed below. The kief is then pressed mechanically or by hand.

The two most popular kinds of hash available nowadays are the ice water hash and those that are chemically extracted. Both types are made using more techniques.

Ice water hash, also called bubble hash, is produced by stirring the cannabis plant into a bucket of water that is cold enough to freeze and take the trichomes off the plant material. The water-weed mixture is then poured into a series of 3 to 5 superfine nylon mesh screens. Once the mud-like trichomes are collected and dried, they are made into hash with 70-80% THC range.

The other technique, as mentioned, involves chemicals. THC glands aren't soluble in water, but they dissolve in chemicals like rubbing alcohol and butane. For extractions using butane, cannabis is crammed in a long glass or metal cylinder with both ends open. The liquid butane is then sprayed into the tube and down the bottom of the cylinder where it is filtered.

This technique allows the THC to dissolve into the butane. It is then collected, carefully heated and placed in a vacuum chamber for evaporation. After which, the potent THC wax (aka budder or BHO) is produced, clocking in at 90% THC or even higher.

A bowl with hash

What is ABV?

Already been vaped (ABV), also called AVB (already vaped bud), is the plant matter left over after you vaporize marijuana. It usually has a dark-brown color and a distinct, subtle aroma in comparison to fresh buds. Many people simply toss their ABV in the garbage or ashtray, thinking they already got all the essence and mood-enhancing benefits out of it. However, there are still many ways to get the most out of your ABV.

Since ABV has already been de-carbed, you can use it directly for the following:

Smoke - You can still smoke cannabis that has already been vaped, although not a lot of marijuana users do this. ABV tastes bad and does not deliver as much kick as fresh buds do. People generally opt to vaporize marijuana because it does not combust the herb, so smoking ABV totally beats this purpose. Anyhow, you can still get a slight punch from smoking your ABV.

Capsules - You can fill empty gel capsules with your ABV buds. This is a good and popular method to ingest ABV cannabis oil because it's simple, does not taste bad and allows you to regulate your doses better.

ABV coconut oil - This works great for snacks, salad dressings and pastas. Combine coconut oil and ABV in a saucepan over low heat until they are incorporated. You may use soy lecithin to bind everything together. Strain and store the oil in the refrigerator until ready for use.

Cannabutter - Cannabutter is a popular ingredient in so many cannabis-infused recipes. Grind the ABV buds in a coffee grinder until fine. Combine water, unsalted butter, and your ABV in a saucepan over low heat and simmer for a few hours. Store in the refrigerator until the butter hardens on top.

Direct approach - If you're down with ingesting the plant material directly, you can sprinkle a few grams of ABV buds onto your sandwiches, pizzas and pastas, or into your drinks and smoothies. ABV tastes much like fresh buds with slightly the same aroma and a little more "toasted" feature.

Already been vaped marijuana (ABV)

Why and how to decarboxylate the buds

When raw, the cannabis plant is non-psychoactive. It only becomes so when the buds age and dry out or when they are heated. Heating creates more psychoactive compounds than ageing. To release the maximum potential of the plant's psychoactive properties, the decarboxylating or "decarbing" process must be performed.

This process is only necessary when you're cooking recipes with cannabis or when you are creating topical solutions and tinctures. Decarboxylating is not necessary when vaporizing or smoking cannabis; the high temperatures decarb the weed automatically.

Here's how to decarboxylate cannabis buds:

1. Preheat oven to 240° F.

2. Break up cannabis buds and flowers into smaller bits using your hands.

3. Place the broken pieces in a single layer on a baking pan lined with parchment paper. Make sure that the pan is the right size so there's no empty space on the sheet.

4. Bake the buds for 30-40 minutes. Stir every 10 minutes for even toasting.

5. When the buds have dried out and turned darker in color, remove from oven and allow the buds to cool. The texture should be crumbly when held.

6. Pulse the buds in a food processor until coarsely ground, but not too fine and powdery.

7. Store in a glass, airtight container until ready for use.

The same process can be done for decarbing trim, kief and stems.

Marijuana ready for the oven

Infusing Cannabis

When infusing cannabis, choose an oil with high fat content and that is made from a natural, unmodified crop. Avoid using vegetable and canola oil and always go for the unrefined ones. Below are the best oils to use for cannabis infusion:

Coconut Oil - Coconut oil is one of the best choices for cannabis extractions because it has the highest saturated fat content. It is capable of absorbing more cannabinoids than other oils. It also has better shelf life when heated.

Olive Oil - Olive oil is incredibly flavorful and healthy. It works perfectly for salad dressings, pastas, flatbread pizza or as a dip for your warm morning bread. Tip: Use high-quality virgin oil made from real, natural olive oil and not the repackaged or colored versions of vegetable oils.

Red Palm Oil - Although tasty, red palm oil is seldom used in American recipes. The oil itself is good for frying over high temperatures, but it can cause the more delicate cannabinoids and terpenes disappear. As such, it is best to use cannabis-infused red palm oil at low temperatures only or as an addition to curry and stir fry at the end.

Walnut Oil - Walnut oil works best for cannabis-infused bruschetta, pesto, and other savory Italian recipes. However, it is usually pricier than other types of oil. You can blend equal parts of MCT oil (Medium-Chain Triglyceride) and walnut oil to make it cheaper and more potent.

Butter - It is best to use unsalted, organic butter for cannabis extractions. You can replace regular butter with canna-butter to make special brownies, space cakes, or as a spread for your morning toast.

Infusing these oils with cannabis is really easy. You can follow the same extraction method you would use for making cannabis-coconut oil. Simply replace coconut oil with your preferred type of oil.

Safety Precautions

1. Select a large area with good ventilation to make cannabis-infused oils. Open the windows and have at least one working fan for excellent air circulation. If possible, make your oils outside in the open air to lower risks of fire or explosions.

2. If you have respiratory problems or sensitive lungs, wear a facemask especially if you're using alcohol. This is to protect your air passages from the fumes when the alcohol is evaporating.

3. Be sure to use proper setup equipment. If you're doing an infusion that involves solvents, use a double boiler type pan and pot setup, with solvent pan sitting on top of the water pot, and not directly on the heating element of the stove. This is applicable to both gas and electric stoves.

4. Take extra caution when you're making oil or butter on a gas stove. Never place the solvent pan near the stove when the flame is on as it can catch flame quite easily.

5. Make your oil on your most convenient time, when there are no distractions and you're not in a hurry. Getting distracted or doing things in a hurry can result to unexpected hazards.

6. Once you're done making your canna oil or butter, make sure to turn off the stove, oven and any other open-flame heating element that you have used.

7. Store your oils and edibles in child-resistant containers and in a dry place where it can't be reached by children and pets. Accidental ingestion of cannabis and cannabis-infused edibles by children and pets may cause health problems.

8. Oil is a highly flammable material, so it is a must to follow these preventative measures for optimal safety. In case of fire, do not panic and follow proper fire extinguishing procedures. Contact the fire hotline if necessary.

Marijuana Recipes

Using infusion

Making Cannabis Oil (coconut oil)

What do I need?

- ✓ Small saucepan or crockpot
- ✓ Wooden spatula
- ✓ Metal strainer
- ✓ Cheesecloth
- ✓ Large measuring cup or bowl
- ✓ Glass jar/bowl for storage

Ingredients

- • 2 cups unrefined coconut oil
- • 40 grams of decarboxylated cannabis

How to make it

1. Combine coconut oil and cannabis in a small saucepan (or crockpot) over minimum heat. Let the oil melt then simmer for one hour. Do not cover the pan. Stir frequently for even cooking.

2. Place the metal strainer and cheesecloth over the measuring cup/bowl. The cheesecloth will filter out the finest particulates that escape the strainer.

3. Pour the cannabis and oil mixture into the cheesecloth and let it drip for about an hour. Wrap the plant material with the cheesecloth and squeeze all the remaining liquid out. You may throw the leftover plant material to the compost or mix it with butter to use as spread, although it will no longer have much effect since most of the plant's essence have been squeezed out.

4. Let the mixture cool for a few minutes and let it solidify. Once it reaches room temperature, close the glass container and keep it in the refrigerator. This recipe is usually good for up to one year.

Cannabis Butter

Cannabis butter from the store

What do I need?

- ✓ Medium saucepan or stockpot
- ✓ Spatula
- ✓ Cheesecloth or metal strainer
- ✓ Spoon
- ✓ Glass container

Ingredients

- 1 ounce of cannabis flower, ground
- 1 cup of water
- 1 lb. unsalted butter

How to make it

1. Add water and unsalted butter into a saucepan or stock pot over medium heat. Let the butter completely melt into the water while it simmers on low heat. You may add water to regulate the temperature and prevent the butter from burning.

2. Add the ground cannabis flower gradually and stir as you do so. Simmer for 2 to 3 hours, stirring occasionally. Be sure the mixture does not come to a boil.

3. After simmering, pour the mixture into a glass container with tight-fitting lid. Use fine metal strainer or cheese cloth to filter out the plant material from the butter mix.

4. Squeeze the cheese cloth or use the back of a spoon to get all the liquid off the plant. Discard residual plant material.

5. Let the mixture cool for a few minutes. Cover the container and store in the refrigerator overnight or until the cannabis butter has fully hardened.

6. The hardened butter will separate itself from the water, making it easier to take out the canna-butter and add to your recipes.

7. Dispose the remaining water properly. Store the canna-butter at room temperature until it's ready for use.

Cannabis Brownies

A big cannabis brownie

What do I need?

- ✓ Mixing bowl
- ✓ Spatula
- ✓ Oven

Ingredients

- • 1/3 cup canna oil
- • 1/3 cup water
- • Brownie mix
- • ½ cup chocolate chips (semi-sweet)
- • 2 large eggs, beaten
- • Cooking spray
- • Melted chocolate

How to make it

1. Preheat oven to 340° F. Grease a 9 x 13-inch brownie pan using the cooking spray.

2. In a large bowl, combine canna oil, water and eggs. Add the brownie mix gradually and mix until the ingredients are completely incorporated. Add the semi-sweet chocolate chips.

3. Pour the batter into the prepared brownie pan. Spread evenly.

4. Refer to the baking instructions found on the box of the brownie mix. To check whether it's already cooked, insert a toothpick into the brownies. If the toothpick comes out clean, proceed to next step.

5. Remove the brownies from the oven and let them cool for half an hour. Cut into equal pieces.

6. Place the chocolate in a medium bowl and melt it in the microwave for 20 seconds. Take out and stir well. Heat it again in the oven for 10 seconds. With a wooden spoon or spatula, stir the chocolate briskly until creamy and smooth.

7. Dip the brownies in the chocolate and coat evenly. Top the coated brownies with nuts or pecans, if desired. Serve warm and enjoy.

Cannabis Granola Bar

Cannabis infused granola bar

What do I need?

- ✓ Mixing bowl
- ✓ Spatula
- ✓ Oven for baking

Ingredients

- 2/3 cup melted canna-butter
- 2/3 cup honey
- 2 teaspoons vanilla extract
- 1 teaspoon baking soda
- 1/3 cup brown sugar
- 1 cup all-purpose flour
- 4 to 5 cups oatmeal
- Coconut oil spray
- 2-cup mixture of coconut flakes, raisins, pecans, walnuts, and - Chocolate chips (you may decide on the ratio)

How to make it

1. Preheat oven to 325° F. Grease a 9 x 13-inch baking pan with the coconut oil spray.

2. Combine all ingredients in a large bowl. Mix well.

3. Spread the mix evenly into the prepared baking pan and press it by hand. Be sure that the thickness is even.

4. Bake for about 30 minutes, or until golden brown.

5. Remove and let cool for 30 minutes. Cut into equal pieces.

6. Serve warm and enjoy.

Cannabis Peanut Butter Cookies

What do I need?

- ✓ 2 mixing bowls
- ✓ Wooden spatula or spoon
- ✓ Microwave oven

Ingredients

- ½ cup canna-butter
- ½ cup peanut butter
- 1 cup white sugar
- ½ cup brown sugar
- 1 teaspoon vanilla extract
- ½ teaspoon baking powder
- ½ teaspoon baking soda
- 1¼ cups flour
- ½ cup white sugar (separate serving)
- 1 egg

How to make it

1. Preheat oven to 350° F.

2. Microwave the canna-butter for 30 to 50 seconds until mostly melted.

3. Mix the melted canna-butter and peanut butter together. Add in half of the flour.

4. Add white sugar, egg, brown sugar, baking powder, baking soda and vanilla extract. Mix thoroughly. Add the remaining flour.

5. In a separate bowl, pour the other serving of white sugar.

6. Shape the dough into small balls using a spoon. Make 24 balls.

7. Roll the dough balls across the bowl of sugar until they are thoroughly coated.

8. Place the coated cookie balls on a baking sheet. Bake for 7 to 10 minutes, or until the cookies harden and start to crack on top.

9. Take out the cookies and let cool. Serve warm and pair with a glass of milk.

Cannabis Chocolate Dipped Fruit

What do I need?

- ✓ Mixing bowl
- ✓ Spatula
- ✓ Microwave oven

Ingredients

- 2 tablespoons of cannabis-coconut oil
- 1 ½ cups of chocolate chips
- 12 fresh strawberries

How to make it

1. In a medium mixing bowl, combine coconut oil and chocolate chips.

2. Set the microwave on high heat and bake the mixture for 30 seconds.

3. Take out and stir. Heat it again for 15 seconds and stir. Repeat this process until the mixture is smooth and completely melted.

4. Remove and let it cool down to room temperature.

5. Dip the fresh strawberries into the chocolate and set them nicely on parchment paper. Let sit for 30 minutes and serve. Enjoy!

Cannabis Tea

What do I need?

- ✓ Your favorite teacup

Ingredients

- 1 teabag
- 1 teaspoon canna-butter
- Milk (optional)

How to make it

1. Add the canna-butter and teabag to a teacup.

2. Pour warm/hot water.

3. Stir and allow the butter to dissolve completely.

4. Take out the teabag and add milk. Consume while still warm.

Cannabis Caramel Sauce

Lovely caramel sauce

What do I need?

- ✓ Small saucepan
- ✓ Spatula
- ✓ Glass, airtight jar for storage

Ingredients

- ½ cup canna-butter
- ¾ cup heavy cream
- ½ cup granulated sugar
- 1 cup brown sugar
- ½ tbsp. vanilla extract
- ½ teaspoon sea salt
- ½ cup distilled water

How to make it

1. In a small saucepan over medium heat, mix the canna-butter, water, salt, brown sugar, heavy cream and granulated sugar together.

2. Whisk the mixture gently and constantly for 8 minutes, or until it starts to thicken. Continue whisking to prevent the sauce from burning. Adjust the heat setting as necessary.

3. Add vanilla extract and stir. Heat for another minute.

4. Remove from heat and let the sauce cool for about 5 minutes.

5. Transfer the cannabis caramel sauce to a glass jar with lid. Refrigerate until cold. Store until ready for use.

Cannabis Cookies

What do I need?

- ✓ 2 mixing bowls
- ✓ Small saucepan
- ✓ Spatula
- ✓ Microwave oven
- ✓ Glass container with lid

Ingredients

- 3/4 cup canna-butter
- ½ cup white sugar
- 1 cup dark brown sugar
- 2 large eggs
- ½ teaspoon fine salt
- ¾ teaspoon baking soda
- 2 ¼ cups all-purpose flour
- 1 tbsp. vanilla extract
- 12 ounces (1 bag) semi-sweet chocolate chips
- ½ cup pretzels or 1 cup walnuts (optional)

How to make it

1. Preheat oven to 375° F.

2. Line 2 cookie sheets or grease them with regular butter.

3. Melt the canna-butter in a small saucepan and let it cool down for a few minutes. Alternatively, put the canna-butter in a microwave-safe bowl and heat in the oven for 30 seconds.

4. In a large bowl, combine sugars, eggs and vanilla extract. Add the melted canna-butter and stir until the mixture is smooth.

5. In a separate bowl, combine salt, baking soda and all-purpose flour. Mix until the ingredients are completely incorporated.

6. Mix wet mixture from the first bowl with the dry mixture from the other. Do not over-mix. Add the chocolate chips.

7. Add the pretzels or walnuts, if using.

8. Scoop about a quarter cup of the cookie dough and place on the prepared baking sheets. Set the cookies about 2-3 inches apart.

9. Bake for 12-16 minutes. Remove from the oven and let cool for a few minutes.

10. Store the cannabis cookies in a glass container with lid. Consume within 5 days.

Cannabis Sugar

Cannabis sugar

What do I need?

- ✓ Mixing bowl
- ✓ Spatula
- ✓ Microwave oven
- ✓ Fine metal strainer
- ✓ Glass container w/ tight-fitting lid

Ingredients

- 1 cup granulated sugar
- 1 tbsp. alcohol-based cannabis tincture

How to make it

1. In a mixing bowl, combine granulated sugar and cannabis tincture. Mix well. The texture should be grainy.

2. Spread the mixture on a baking pan lined with baking sheets. Heat it in the oven at 200° for 30 minutes.

3. Remove from the oven and let it cool for some minutes.

4. Prepare an airtight glass container and put a metal strainer or sieve over it. Place the clumped sugar in the strainer and press it using the back of a metal spoon. This will break up the sugar and create a smooth consistency.

5. Continue pressing the sugar into the strainer until everything is broken up.

6. Seal the container and store in a cool, dry place until ready for use.

Cannabis Chocolate Ice Cream

What do I need?

- ✓ Medium saucepan
- ✓ Spatula
- ✓ Cheesecloth
- ✓ Mixing bowl
- ✓ Glass bowl for chilling
- ✓ Ice cream scoop

Ingredients

- 2 tbsp. cannabis-coconut oil
- 1 cup milk
- ¾ cup sugar
- ¼ teaspoon salt
- 2 tbsp. cocoa powder, unsweetened
- 3 eggs, lightly beaten
- 2 ounces semi-sweet chocolate, chopped
- 1 teaspoon vanilla extract
- 2 cups heavy cream

How to make it

1. In a medium saucepan, combine cannabis oil, sugar, salt, milk, and cocoa powder. Cook over medium heat until smoothly blended. Stir constantly.

2. Remove mixture from heat and filter it with a cheesecloth. Take all the liquid out and pour it back to the saucepan. Bring the mixture to a simmer.

3. In a small bowl, whisk the egg yolks while gradually adding about ½ cup of the mixture from the saucepan. Return the mixture with the egg yolks to the saucepan, whisking constantly.

4. Heat until the mixture thickens. Do not boil. Adjust heat if necessary.

5. Remove from heat and add chopped chocolate. Stir until the chocolate melts.

6. Transfer the chocolate mixture in a bowl for chilling. Refrigerate for about 2 hours, stirring occasionally.

7. When the mixture has completely cooled, add heavy cream and vanilla extract.

8. Pour the mixture into an ice cream maker and freeze until the ice cream is firm.

9. Let the ice cream stand for 5 minutes at room temperature before scooping. Serve, share and enjoy.

Using ground marijuana

Space Cake

Heavenly space cake

What do I need?

- ✓ Coffee grinder
- ✓ 2 mixing bowls
- ✓ Spatula
- ✓ Oven for baking

Ingredients

- • 8 grams cannabis buds
- • 1 cup unsalted butter
- • 1 cup granulated sugar

- 1¾ cups all-purpose flour
- ¾ cup milk
- ½ teaspoon of salt
- 1¾ teaspoons of baking powder
- 2 eggs

How to make it

1. Grind the cannabis buds in a grinder (coffee grinder works great) until it is fine and consistent.

2. Preheat oven to 190° C.

3. Place the butter in a small bowl and heat in the oven for 20 seconds until it melts into a pasty consistency.

4. In a medium bowl, mix the melted butter with the ground cannabis. Add in sugar, milk, flour, eggs and salt. Mix until the batter has an even consistency. If the mixture is too dry, splash extra milk. If it is too wet, put in a little more flour.

5. Line the baking pan with baking sheets or grease it with non-stick spray.

6. Pour the batter into the pan and spread evenly. Bake for 25 minutes.

7. Insert a skewer into the center of the cake to check if it's ready. If the skewer comes out clean, proceed to next step. If not, heat the cake for another 5 minutes.

8. Remove from the oven and let the cake cool for about 20 minutes. Place the space cake upside-down on a wire rack. Leave it for another 30 minutes to cool.

9. Set the cake on a platter and add some decoration if you wish. Serve and enjoy!

Cannabis Pizza

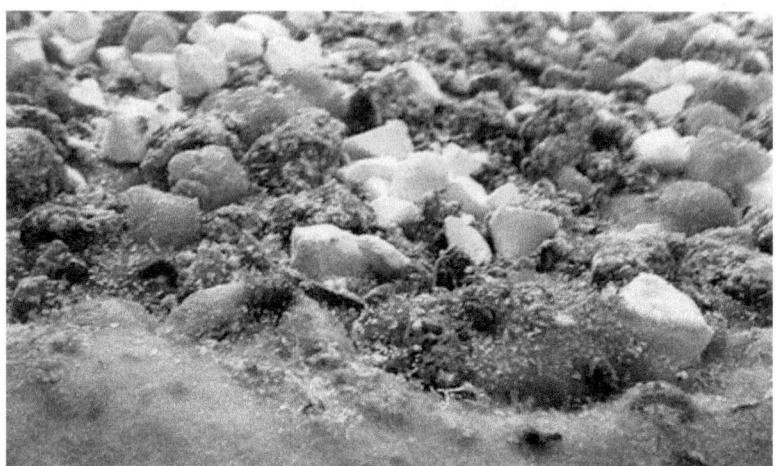

Pizza with cannabis on top

What do I need?

- ✓ Mixing bowl
- ✓ Rolling pin
- ✓ Spatula
- ✓ Saucepan
- ✓ Ladle
- ✓ Oven for baking

Ingredients

- ½ cup canna oil
- 200 grams flour
- 5 grams fresh baking yeast
- 2 large tomatoes, cut up
- ½ tbsp. sugar
- ½ teaspoon pepper
- ½ teaspoon salt
- ½ tbsp. dried oregano
- ½ tbsp. dried rosemary

- 1 glass of warm water
- 1 to 2 tbsp. cannabis buds, ground
- Individual toppings (e.g. mushrooms, onions, bacon, salami, mozzarella)

How to make it

The Dough:

For the dough, you may either make it yourself or buy a prepared one from a store. Tip: Creating the dough yourself will make the pizza taste better.

1. Mix the baking yeast in a glass of warm water. Add sugar and a pinch of salt. Let stand for 10 to 15 minutes.

2. In a mixing bowl, combine the flour with a pinch of salt and 4 tablespoons of canna oil.

3. Add yeast and knead for 10 minutes, or until the dough non-sticky soft and elastic.

4. Once the dough is ready, cover it and store in a dry, warm place for about an hour.

The Sauce:

1. In a saucepan, combine tomatoes with pepper, rosemary, oregano, some salt, and ½ cup of warm water. Simmer for 15 minutes over medium heat.

2. The amount of each seasoning depends on your preference. Give the sauce a taste until you find the perfect balance. If the sauce becomes too low, add a little more water. If it's too watery, just let it boil away.

The Pizza:

1. Roll the prepared dough and pour the tomato sauce onto it. Spread evenly. This is your basic cannabis pizza.

2. Add toppings to your liking. Popular topping choices for pizza are mozzarella cheese, bacon, onions, salami and mushrooms. You can also add pineapple, tuna, shrimps or asparagus for more flavor.

3. Bake at 200° C for 18 to 20 minutes, until the dough becomes brown.

4. Once the pizza is cooked, sprinkle ground cannabis buds over it for a more intense effect and to make the presence of the cannabis more visible. You can also drizzle it with cannabis oil.

5. Serve hot and enjoy!

Cannabis Muffins

Cannabis muffins

What do I need?

- ✓ Saucepan
- ✓ Spatula
- ✓ Metal strainer or cheesecloth
- ✓ 2 mixing bowls
- ✓ Muffin cups
- ✓ Oven for baking

Ingredients

- 6 grams ground cannabis
- ½ cup 100% vegetable oil
- 1 tbsp. unsalted butter
- 1/8 teaspoon cinnamon, ground
- 2 tbsp. all-purpose flour
- 1/3 cup brown sugar, packed
- 1 egg, beaten
- ¾ cup white sugar
- 3 mashed bananas or 1 cup of fresh strawberries
- ½ teaspoon salt
- 1 teaspoon baking powder
- 1 teaspoon baking soda
- 1½ cups all-purpose flour

How to make it

1. Pour vegetable oil into a saucepan and add the ground cannabis. Simmer for about 20 minutes over medium heat.

2. Remove the mixture from the pan and let it sit for a few minutes until it cools.

3. Strain the mixture with a metal strainer or cheesecloth and squeeze all the oil out. Discard the leftover plant material and set the extracted oil aside.

4. Preheat oven to 375° F.

5. Line 10 muffin cups with muffin papers or lightly grease with melted, unsalted butter.

6. In a large mixing bowl, combine flour (1 ½ cups), baking powder, baking soda, and salt.

7. In a separate bowl, combine egg, sugar, bananas (or strawberries), and the extracted oil. Mix well.

8. Combine the banana/strawberry mixture with the flour mixture until the batter moistens. Fill about 3 quarters of the prepared muffin cups with the batter.

9. In a small mixing bowl, combine brown sugar, cinnamon, and 2 tablespoons of all-purpose flour. Add 1 tablespoon of butter until the mixture resembles a coarse cornmeal. Top the muffins with this mixture.

10. Bake for about 20 minutes. Insert a toothpick into the center of the muffins to check if they are ready. If the toothpick comes out clean, remove the muffins from the oven. If the batter sticks to it, bake for another 5 minutes.

11. Set on a plate and serve.

Using Decarbed Marijuana

Cannabis Chicken Noodle Soup

Fresh cannabis noodles

What do I need?

- ✓ Food mixer (a vegetable chopper works fine, too)
- ✓ Pizza cutter & rolling pin (or noodle maker, if you have one)
- ✓ Large crockpot
- ✓ Spatula

Ingredients

- 1 cup decarbed cannabis
- 4 cups all-purpose flour
- 3 eggs
- 14.5 ounces vegetable broth
- 60 ounces chicken broth
- ½ pound cooked chicken breast, chopped
- 1 tbsp. canna-butter
- ¼ cup peas
- 1 cup carrots, sliced
- ½ cup onion, sliced
- ½ teaspoon oregano, dried
- ½ teaspoon basil, dried
- ½ cup carrots, chopped
- ½ cup celery, chopped
- Salt and pepper

How to make it

The noodles:

1. In a food mixer or vegetable chopper, combine the decarbed cannabis buds with flour and eggs. Make sure the ingredients are completely incorporated. You should have a sticky cannabis dough as a result.

2. Lay the cannabis dough on a clean surface and start kneading it with the rolling pin. Maintain the thickness as even as you possibly can. Start cutting the dough into strips using the pizza cutter. If you're using a noodle maker, follow manufacturer's instructions for making the noodles.

3. Set the noodles aside or keep it in the fridge until ready for serving.

The noodle soup:

1. Melt the canna-butter in a large crockpot over medium heat. Be sure not to bring it to a boil.

2. Cook the celery and onion in the butter for about 5 minutes. Add vegetable and chicken broths, stirring frequently.

3. Proceed to add carrots, oregano, chicken and basil. Sprinkle salt and pepper to taste. If you want to add more kick to your soup, you may add about 6 droppers of cannabis tincture. Bring to a low boil.

4. Once the bubbles start to appear, lower the heat and simmer for about 20 minutes. Add in your prepared cannabis noodles about 15 minutes before serving.

5. Remove from heat and serve hot. Enjoy!

Using ABV

ABV Coconut Oil

ABV coconut oil

What do I need?

- ✓ Mason jar
- ✓ Slow cooker
- ✓ Metal strainer
- ✓ Glass container for storage

Ingredients

- 1 cup ABV, finely ground
- 1 ¼ cup coconut oil

How to make it

1. Pour the ABV into a mason jar and add all the coconut oil.

2. Add a few cups of water to a slow cooker and put the filled jar inside.

3. Set the slow cooker on "keep warm" and leave the coconut oil and ABV to cook for 5 hours.

4. Take out the jar from the slow cooker. Place a metal strainer on a glass storage container and start straining the contents of the mason jar. Use a spoon or the bottom of the jar to press the ABV and take all the oil out.

5. Repeat the straining process if you can still see some particulates in the oil. Use a second, finer strainer if necessary.

6. Once the liquid has cooled, keep it in the refrigerator until it hardens.

7. Store properly until ready for use.

The End

This book has come to an end. Do not expect to be a great marijuana grower and cook already. There is still so much to learn. By now you should have a good understanding of how it works and how you can troubleshoot your plants.

Learning is not over yet! The part where you will learn the most is when you plant them yourself and refer to this book for advice. Also, you should participate on marijuana growing forums and Facebook groups to get the most out of your experience.

Thank you for reading this book.
It has been a pleasure to write it. Can I ask you a favor? If you can spare a few minutes, could you please leave a review about this book? It helps me to produce even more books about this topic.

Good luck growing!

www.ingramcontent.com/pod-product-compliance
Lightning Source LLC
Chambersburg PA
CBHW051545170526
45165CB00002B/884